BE AN
ADVANCED DRIVER

BE AN
ADVANCED DRIVER

by

Jennifer D. Robbins
D. Tp. A.D.I. M.I.A.M. Dip.D.I.

Peter W. Robbins
D. Tp. A.D.I. M.I.A.M. Dip.D.I H.G.V.I.

Shaw & Sons Ltd
Shaway House
London SE26 5AE

Published in May 1989 by Shaw
& Sons Ltd of Shaway House,
London SE26 5AE

ISBN 07219 1043 2

Text typeset by MacLink, Grimsby
and printed by the Gordon Press,
Croydon.

Contents

Windscreen and windows

Foreword

The authors of this book have a total of 40 years' driving experience and run two driving schools in Ross-on-Wye, Herefordshire. Both are Department of Transport Approved Driving Instructors and members of the institute of Advanced Motorists. They hold an additional qualification, the Diploma of Driving Instruction which is awarded jointly by the Associated Examining Board and the Driving Instructors Association. Only a very small percentage of the driving instructors possess this qualification. Peter Robbin's also holds the coveted D.I.A. Advanced Instructor qualification.

Having already written two popular books for learner drivers entitled, *How To Pass Your Driving Test and Highway Code Questions and Answers*, the authors decided to use their professional driving experience to write a comprehensive reference book for drivers who wished to improve their driving to an advanced standard. The contents include detailed information on all aspects of safe driving ranging from correct motorway procedures to dealing with floods, burst tyres to making an insurance claim against another driver. This is an extremely useful reference book and should be kept in a handy place by all motorists.

Introduction

What makes an advanced driver?

A driver does not become "advanced" just by the length of time he or she has been driving, nor even by the number of miles driven. Experience is obviously both helpful and important, but drivers gain this experience throughout their driving lives. Advanced driving means being safe: it is estimated that at least 90 per cent of all accidents are avoidable and drivers need to be safe all of the time.

Mental outlook is important. A good driver should be patient and courteous towards other road users. Continuous concentration and a sense of responsibility are vital to cope with the stressful situations that occur on overcrowded highways. Above all, drivers should cultivate awareness and anticipation to assess possible dangers *before* they become accidents.

A basic elementary knowledge of car mechanics is helpful, as without it a driver will not be aware of the limitations of the vehicle being driven and therefore not totally in sympathy with the controls of the car. This book emphasises the importance of a vehicle being in good mechanical condition, encourages pre-driving checks and stresses the necessity of having tyres in a legal and roadworthy condition.

Finally, conscientious drivers will want to increase their knowledge of correct road procedures, always aiming to keep their driving at a high standard of competence and applying the Highway Code at all times. Drivers should plan driving actions well in advance, constantly anticipating hazards and new situations before they arise.

Drivers who "read the road ahead" in this way, are

not taken by surprise. They are always travelling at the right speed, in the correct gear, and in an ideal position on the road to react safely when problems occur. Driving is not just getting from A to B, it is *how* you get there that matters. Indeed, some drivers do not get there!

This wide-ranging and informative driving book also deals with events and problems likely to be outside the range of experience of many readers. There are sections on how to avoid aquaplaning, dealing with a burst tyre and driving safely through flood water. This is apart from such useful and practical items on how to make insurance claims, towing a trailer and which tinted lenses are safest to wear when driving. A handy reference book to keep in the car.

Enjoy your driving...and take pride in it.

☞ Whilst every attempt has been made to ensure the complete accuracy of the information given in this book, neither the authors nor the publishers can accept liability for any error or misinterpretation resulting from its use.

Accident procedure

At the scene of an accident

- ☐ STOP in a safe place.
- ☐ Your vehicle must not endanger other traffic.
- ☐ Your vehicle should not cause any obstruction.
- ☐ Allow plenty of room for emergency services if needed.

Warning other traffic

- ☐ Warn traffic quickly without endangering yourself.
- ☐ Use lights, hazard lights, cones or warning triangles.
- ☐ Put out cigarettes or naked lights.
- ☐ Switch off the engine of vehicles involved in the accident.

Warning reflective triangles should be placed on the same side of the road 50 metres before the obstruction or if on a bend, well before it, and 150 metres on a motorway. Cones should be placed 15 metres from the obstruction next to the kerb and subsequent cones should guide traffic safely around the obstruction. After collision, fire is the main danger, especially where petrol has been spilled.

Alerting the emergency services

- ☐ Dial 999 and state which services you require.

- [ ] On a motorway use the nearest emergency telephone.

- [ ] State the location, type of vehicles involved and number of casualties.

Whoever contacts the emergency services must give clear and accurate details. If an accident involves a vehicle displaying hazard information panels, ensure the emergency services are given all the available details, but some goods, for example, poisonous gasses and radioactive loads, require full protective clothing to be worn so *keep well back.*

Dealing with casualties

- [ ] Do not move casualities unless they are in obvious danger.

- [ ] Do not remove a motorcyclist's safety helmet.

- [ ] Keep people warm and reassured.

- [ ] Apply first aid if necessary.

- [ ] Give nothing to eat or drink.

People who are injured are much more likely to go into shock when removed from the protection of their vehicles. Use blankets, coats or jackets to keep them warm and talk quietly reassuring them until the ambulance arrives. Sometimes serious internal injuries may not be immediately apparent, and moving a casualty who has spinal or neck injuries can result in paralysis or death. Only consider moving an injured person if another life threatening danger such as fire presents a serious risk. Never attempt to remove the helmet of a motorcyclist, this may aggravate head or neck injuries.

First aid

☐ Carry a first aid kit in the car.

☐ Learn first aid: St. Johns Ambulance, St. Andrews and the Red Cross run good courses.

The type of life-saving first aid which may be needed at road traffic accidents while waiting for the emergency services to arrive, principally concerns breathing and bleeding.

If breathing has stopped

☐ Clear throat of injured person of tongue, false teeth etc.

☐ Support neck and tilt head back.

☐ If breathing has not restarted, hold nostrils shut and blow into the mouth every four seconds. Watch the chest rise and fall. Stop when breathing starts.

How to stop heavy bleeding

☐ Apply firm pressure to the wound using a clean cloth.

☐ If glass or other sharp material is in the wound, apply the pressure to a pressure point before the wound.

☐ Raise unbroken arms or legs to lessen bleeding.

Procedure for those involved

☐ You must stop as soon as possible.

☐ Give your name, address and car registration number to those involved.

☐ Give the owner's name and address if different from your own. If you are unable to do so, report the accident to the police within 24 hours and produce your insurance certificate.

☐ Report the accident to the police within 24 hours, if another person is injured, or if someone else's property has been damaged.

☐ You must provide details of your insurance following an injury in an accident and to the person who holds you responsible for damage to their property.

These are the steps you must take if there is injury to anybody or damage to other vehicles or property, which includes fences, street lights and road signs. When you report the accident to the police they may want to see other driving documents.

Information needed

☐ Names and addresses of other drivers involved.

☐ Details of the other cars involved.

☐ Insurance details of the other drivers involved.

☐ Names and addresses of passengers and witnesses.

☐ Make a plan or sketch showing the position of the vehicles.

☐ Note the time and exact location of the accident.

☐ Write down names and numbers of the police who attend the accident.

Do not admit liability or say anything that could be used against you at a later stage such as, "I am sorry". Take any name or address which could prove useful later and also note down the registration numbers of cars parked in the vicinity, so that the owners can be traced if necessary.

Notify your insurance company even if you don't intend to claim. As soon as practical, write a full account of what happened and make a sketch showing the position of the vehicles. It may be useful to carry a camera and take a photograph.

Leaving the scene

Wait until the emergency services arrive before leaving the scene of the accident. Even if your vehicle was not involved, the police may wish to question you regarding details of the accident. Do not allow casualties to wander off, as they may be concussed or in shock.

Insurance claim against another driver

☐ Contact your insurance company and ask for a motor vehicle accident claim form.

☐ Complete the form and return it promptly.

☐ State clearly if you wish a claim to be made against the insurance company of another driver.

☐ Do not allow any damage to your vehicle to be repaired until the insurance company assessor has seen it.

☐ Claims for expenses should be made direct to the other insurance company as soon as is practical.

It is a good idea to keep a claim form at home. Fill it in in pencil first in case of errors. "No claims" bonuses are *no claims* not *no blame* bonuses. Liability must be proved against another driver before their insurance company will pay out. If delays occur, ring your company, take the name of the person you deal with and ask for the claim to be attended to promptly.

Expenses such as car hire, any excess you may have had to pay, damage to personal property etc. should be claimed from the other insurance company direct. Send your claim together with relevant receipts and telephone if there is any undue delay. For large claims or claims for injuries suffered, it may be wise to seek advice from a solicitor.

Bends

Approaching a bend

☐ Slow down: taking account of road, weather and traffic conditions.

☐ Select the correct gear before the bend.

☐ Position your vehicle well to the left of the road.

☐ At night, dip lights for oncoming traffic earlier on left-hand bends.

Your correct speed will depend on visibility, the conditions and the angle of the bend. Speed should be at its lowest the moment you begin to steer into the corner. Make sure you are in a low gear that will give maximum control for the vehicle you are driving. Entering a bend too fast is not only dangerous and more likely to lead to a skid, but will cause more wear and tear to the tyres, steering mechanism and suspension of your vehicle. Keeping to the left entering bends may reduce your field of vision, but is the only safe position should there be an unexpected obstruction ahead or oncoming traffic positioned in the centre of the road. The car should be just under acceleration as you take the corner and speed should reflect the possibility of the hazards mentioned above.

Leaving a bend

☐ As you leave the bend gently increase the acceleration.

Checks prior to driving

Daily checklist

☐ Visually, check the exterior of the car.

☐ Tyres.

☐ Windscreen wash bottle, washers and wipers.

☐ Oil level.

☐ Windscreen and windows.

☐ Lights and indicators.

☐ Loads.

Good drivers allow time to check their vehicle before driving each day. It will only take a few moments. Walk round and check the exterior of the car and the tyres are not damaged. Ensure windscreen washers and wipers are working and that the wash bottle does not need topping up. Check the oil level is correct. Clean all windows and the windscreen, especially when it is icy or has been snowing. Make sure lights and indicators are working properly and are clean, and that a roof rack and any load is secure.

Weekly checklist

☐ Tyre pressures.

☐ Brake fluid and coolant level.

☐ Battery.

☐ Fan belt.

☐ Wheels.

Once a week give your car a thorough check, it could save your life. Check the tyre pressures when cold, including the spare. Lower pressure in a tyre could indicate damage or a puncture. Check the tread depth. The legal minimum is one millimetre, but ideally do not let the tread wear below three millimetres if you wish to stop safely in bad weather conditions. The coolant level should also be checked. Ensure the brake fluid level has not dropped. Top up the battery with distilled water if necessary and check terminals are clean and greased. The fan belt should be looked at for tension and wear. Make sure there is no play in the wheels and the bolts are secure.

Checks prior to moving off

☐ Doors.

☐ Seating.

- ☐ Mirrors.
- ☐ Seat belts.
- ☐ Handbrake and gears.
- ☐ Instrument readings including warning lights and fuel level.
- ☐ Running brake test as soon as possible.

Make sure the doors are shut properly. Adjust your seating so you can reach all controls comfortably. Ensure all the mirrors are clean and correctly adjusted to see behind and alongside the car. Seat belts should be in good condition, and must by law be fastened for driving.

Before starting the engine, check the handbrake is on and the gear lever in the neutral position. When the engine is running, check the warning lights are working properly and there is plenty of fuel for your journey. Finally, as soon as possible do a running brake test to satisfy yourself the brakes are in good working order.

All the above are very basic checks. It is obviously necessary to have your vehicle serviced as recommended in the manufacturer's handbook, and repaired when necessary. Make sure any work is either carried out by an experienced mechanic or supervised by one.

Distractions

Inside the car

- ☐ Passengers talking.

- ☐ Fiddling with the radio or cassette player.
- ☐ Lighting a cigarette.
- ☐ Eating or unwrapping sweets.
- ☐ Opening windows.
- ☐ Children fighting.
- ☐ Looking at maps.
- ☐ Answering a car phone.

The above are just some of the distractions which may occur during driving. Advice regarding car phones is given in the Highway Code: pull over in a safe place to take or make calls. Concentrate on your driving at all times, not the unpleasant news you received that morning. A few seconds' lapse in concentration may be all that is needed to cause a serious accident.

Outside the car

- ☐ Friends or relatives on the pavement.
- ☐ Beautiful scenery.
- ☐ Hot air balloons or planes taking off from an airport.
- ☐ Direction signs.
- ☐ Shop windows.
- ☐ Leggy girls in mini-skirts.

The list is endless, with different drivers being distracted in different ways. Taking your eyes off the road can be fatal, especially when driving at speed or in busy traffic. Concentrate only on your driving.

Eyesight and fitness to drive

Conditions when drivers must not drive

- ☐ When eyesight is defective and uncorrected.
- ☐ When a relevant disability is present.
- ☐ When under the influence of alcohol.
- ☐ When drugs are being taken that could affect driving.
- ☐ When tired or unwell.

A relevant disability is one which would affect driving, for example, attacks of epilepsy. Licence holders must notify the D.V.L.C.. Regarding alcohol, *do not* drink and drive. Driving decisions, driver's reactions and judgement can be seriously impaired *below* the legal limit of 35 microgrammes of alcohol per 100 millilitres of breath. Many drugs that are in common use and can be bought without a perscription can affect driving. Some common examples are cough mixtures, preparation for colds and flu and hayfever remedies. If in doubt check with a pharmacist. When a driver is unwell or tired, reactions will be slow. Do not drive unless you are in good health and feeling fit.

Legal eyesight requirements

A driver must be able to read a modern number plate at a distance of 67 feet, with glasses or contact lenses if worn.

Types of frames suitable for driving

☐ Thin rims and frames with good peripheral vision.

☐ Frames specially shaped for drivers.

Drivers who prefer thick rims or heavy weight plastic frames should keep a second pair in the car for driving, which give good front and side vision.

Rules for driving glasses

☐ Clean glasses regularly.

☐ Keep a spare pair in the car in case of breakage or loss.

☐ Have your eyes examined every two years unless otherwise recommended.

Sunglasses and tinted lenses

☐ Only use these during the daytime.

These should not be worn where the visibility is poor or where the road is shaded. Your optician can advise on various tints.

Photochromic lenses

☐ These do not work well inside a car.

☐ Take care entering tunnels or shaded areas.

These lenses automatically darken when exposed to sunlight. In a car they will probably reduce glare sufficiently to aid driving but do not lighten immediately the light is reduced.

Coloured windscreen strips

☐ Do not stick coloured strips across the wind-screen.

Glasses for night driving

☐ Avoid using yellow-tinted or blue-tinted glasses at night.

☐ Keep a clear pair of glasses in the car for night driving.

Night driving glasses should be avoided as their yellow-tinted lenses make it difficult to see cyclists, pedestrians or parked cars in the shadows, possibly with fatal results. The same applies to blue-tinted glasses where there is sodium lighting. All tinted glasses will reduce visibility in the dark.

It is an offence to drive with uncorrect or defective eyesight.

Flood water

Driving through water

☐ Check the depth of the water if possible.

☐ Use 1st gear.

☐ Slip the clutch and keep the engine speed as high as possible.

☐ Maintain a slow steady speed.

Use your common sense when making the decision whether or not to drive through water. When driving through it, drive at a slow speed and avoid making waves which could cause your vehicle to break down. Keep the engine running at a high rate to prevent water entering the exhaust pipe. After driving through water, check your brakes are working properly. If they are not working, drive slowly with your left foot lightly pressing the brake pedal to dry them out, having checked in your mirror that it is safe to do so.

Fog code

Slowing down

☐ Reduce speed so you can stop safely within your range of vision.

☐ Keep a safe distance from traffic ahead.

☐ Speed is deceptive in fog so allow for unexpected hazards.

☐ Watch for parked vehicles, cyclists and pedestrians.

☐ Obey any special speed limits, especially on motorways.

☐ Allow more time for your journey.

Take extra care where fog is patchy or drifting, and driving in fog after dark.

Rules for following other drivers

☐ Following closely behind another vehicle's tail lights gives a false sense of security .

☐ Do not keep up with vehicles in front unless

they are travelling at a safe speed.

☐ If another car is following too close behind you, do not increase speed.

Hanging on to tail lights can be dangerous. If the vehicle disappears from view it may be the fog has thickened, not that the vehicle has increased its speed. If there is an obstruction ahead, the vehicle in front may stop quickly, so leave enough room to stop safely. The vehicle in front will be displacing thick fog making conditions seem clearer than they are.

Allow more time for a journey.

Lights which should be used

☐ Switch on dipped headlights.

☐ Switch on fog lamps where necessary.

☐ Switch on high intensity fog lights at rear of the car.

It is important to see and be seen. Never rely on side lights when it is foggy. Fog lamps, if you have them, may give better visibility in the dark. Use dipped headlights where they give a better view. High intensity fog lights can dazzle other drivers, so only use them when visibility is less than 100 metres. Park off the road in foggy conditions.

How to improve visibility

☐ Do not allow the windscreen to mist up: use the de-mister and windscreen wipers.

☐ Keep windows clear.

☐ Clean lights, indicators and reflectors when possible.

Make sure the fog you see is outside the car and not on your windscreen. Keep all lights and reflectors clean of grime and moisture so they can be seen clearly. Beware of other vehicles, especially lorries, whose rear lights may be obscured by dirt thrown up from the road.

Gears

Correct use of gears

☐ Check the gear lever is in neutral before starting the engine.

☐ Change gear gently: do not force the lever into gear.

☐ Remove your foot from the clutch after changing gear.

☐ Speed up, then change up, accelerating slightly into the change.

☐ Slow down, *then* change down to the gear you need, depressing the accelerator more to affect a smooth change.

☐ Do not take your eyes off the road when changing gear.

☐ Change gear frequently as necessary: do not drive too long in low gears.

☐ Leave the car in gear when parking on a gradient.

Common faults committed by experienced drivers include driving with the left foot resting on the clutch pedal. This results in a slight loss of control of the car and substantially reduces the life of the clutch. Slow down before changing down. It is not necessary to change down one gear at a time. Block changes, for example, changing from 5th to 3rd, or 4th to 2nd means less wear and tear on the clutch and gear box, and more control over the braking and steering systems of the vehicle.

You can stop in any gear.Drivers who use the gear box to slow the car down will spend more on car repairs and are more likely to be involved in accidents. The graveyards are full of drivers who tried to change down instead of slowing down!

Checking the gear lever is in neutral before starting the engine ensures the car will not jerk forward, possibly with disastrous results, and keeping your foot on the clutch when switching on will prolong the life of the battery. Leaving the car in gear when parking on a gradient ensures the vehicle will not roll should the handbrake fail. Drivers who brake inadequately when approaching junctions and coast up to them with their foot on the clutch pedal, may experience a potentially serious lack of control in certain situations.

Hazard lights and warning devices

Use of hazard warning lights

☐ When your vehicle is stationary to warn of an obstruction.

☐ When your vehicle has broken down.

☐ When you have stopped due to an emergency or accident ahead.

☐ When you break down on a motorway, and are stopped on the hard shoulder.

Always move your vehicle off the road if possible and into a safe place. When hazard warning lights are switched on they flash all four indicators to warn other drivers of your presence.

Where to place a warning triangle

☐ 150 metres behind your vehicle on the hard shoulder of a motorway.

☐ 50 metres behind your vehicle on an ordinary road, or further if you have broken down on a bend.

Switch on your hazard lights, then place the reflecting triangle on the same side of the road as the obstruction, facing oncoming traffic.

Traffic cones and pyramids

☐ Use up to four cones or pyramids.

☐ Place them at intervals to guide traffic past the breakdown.

☐ You may use a flashing amber light, but only with a cone, pyramid or warning triangle.

It is now legal for drivers to carry flashing amber lights, cones and pyramids as part of a emergency kit. The main advantages are their visibility and stability, but must be placed carefully. Always switch on hazard lights first. Flashing amber lights must not be used on their own without other warning signs.

Headlights

Condition of headlights

☐ Keep them clean.

☐ Ensure they are properly aligned and do not dazzle other drivers.

☐ Replace faulty bulbs.

Be the first to switch your headlights on at dusk, and the last to switch off at daybreak. Drivers who drive light or neutral coloured cars will not be seen as easily as a driver in a bright car and should therefore switch on earlier.

When to use dipped headlights

☐ When visibility is poor.

☐ When weather conditions are bad.

☐ During mist and rain travelling on dual carriageways or motorways.

☐ In fog.

☐ At night when there are vehicles in front of you, or oncoming traffic.

☐ In a built-up area at night where there is street lighting.

Remember to dip headlights earlier when travelling

round a left-hand bend and traffic is approaching. When other drivers forget to dip their headlights and dazzle you, slow down as much as necessary but do not flash your lights as you may dazzle them and cause an accident. Sometimes drivers driving on dipped headlights dazzle oncoming drivers quite unwittingly through either having incorrectly aligned headlights or a heavy load in, or to the rear of, their vehicle.

When to use main beam headlights

☐ At night.

☐ Out of 30 m.p.h. zones where there is no street lighting.

☐ When you are not following traffic and there are no oncoming vehicles.

When driving on main beam headlights remember to dip them when either you catch up other vehicles, or oncoming traffic approaches.

Horn

Correct use of horn

☐ The horn should be used as a warning, not a rebuke.

☐ The horn should not be used when the vehicle is stationary, unless to avoid an accident.

☐ The horn must not be used on a moving vehicle between 11.30 p.m. and 7.00 a.m. in a built up area.

☐   Do not sound the horn longer than necessary.

The horn should not be used when you are annoyed with another road user, but as a timely warning. A short, sharp toot is often sufficent to let others know you are there. The horn can be a useful warning when overtaking vehicles whose drivers may be unaware of your presence; approaching sharp bends in country lanes; to alert children who are playing; or in situations where you anticipate another driver is unaware of the presence of your vehicle and may move into your path, risking a collision.

Junction procedure

Approaching a road junction

☐   Look well ahead and note traffic signs and lane markings.

☐   Follow the *mirrors-signal-manoeuvre* routine in good time.

☐   Position the vehicle correctly, keeping to the left-hand side when going straight ahead in the absence of signs or markings to the contrary.

☐   Slow down as necessary, then change gear if required.

☐   Watch carefully for obstructions, pedestrians, motorbikes or cyclists.

☐   Stop at STOP junctions, when necessary at traffic lights, and for pedestrians crossing the road.

It is only necessary to stop at Give Way junctions when it is not clear to go. Always watch for pedestrians who are likely to be crossing the road and give way to them.

Emerging from a road junction

☐ Position the car so you can see both ways clearly.

☐ Assess the situation carefully.

☐ Decide whether it is safe to proceed, wait if it is not, or you are unsure.

It may be necessary to slowly edge out into the junction where there are parked cars, or other obstructions or hazards blocking your view. Never emerge until you are absolutely sure it is safe to do so. Do not beckon pedestrians to cross in case you beckon them into a danger you have not anticipated.

A high proportion of accidents happen to vehicles that are turning right. Take care before crossing the path of other traffic and never cut right-hand corners which could possibly place other vehicles in danger of collision with your vehicle. Remember when turning sharply left into a junction, not to swing out but to anticipate that the back of the car will take a short cut. Do not steer too early and avoid mounting the pavement. Use wing mirrors and check blind spots for cyclists before turning. Always check both ways before driving through or emerging from any junction *whether or not it is your priority.* The art of safe driving is *not* only to avoid causing accidents, but to be able to avoid becoming involved in accidents that other drivers may create.

Roundabout procedure

☐ Look at the road signs and decide which exit you need.

☐ Check in the mirrors.

☐ Signal if necessary.

☐ Select the correct lane in good time.

☐ Follow the *brake - gear - accelerate* routine.

☐ Give way to traffic from the immediate right unless road markings indicate otherwise.

☐ Enter quickly, checking exit roads as you pass.

☐ When ready, use mirrors then signal left to leave. Check blindspots.

☐ If safe, leave in the left-hand lane unless it is obstructed, for example by slow moving traffic.

Roundabouts are junctions where traffic mixes at speed. An accurate judgement of speed and good all-round vision are necessary to enter, circumnavigate and leave safely. Many drivers are not aware of correct signalling and positioning procedures, and an advanced driver is always on guard, anticipating mistakes others might make. If the roundabout is clear, take the shortest possible route through it.

How to signal and position correctly

☐ Turning left: signal left and approach in the left-hand lane.

☐ Following the road ahead: no signal, keep in left-hand lane unless road markings or signs instruct you otherwise. Signal left to leave.

☐ Turning right: signal right and approach in the right-hand lane. Signal left to leave.

Generally speaking where there are several exits, treat the first exit as the left-hand one and signal left on approach. For other exits, keep in the left-hand lane and signal left only to leave, unless they are past the twelve o'clock or straight-ahead position as you approach. In this case, signal right and approach in the right-hand lane. Sometimes arrows will mark the correct lane to be taken, and place names may be similarly marked on the road. When your vehicle passes the exit before the one you are taking, switch on the left-hand indicator to signal your intention to leave the roundabout informing drivers that may be following or waiting to enter.

Roundabout hazards

☐ Cyclists and motorcyclists trying to leave a roundabout.

☐ Articulated lorries negotiating roundabouts.

☐ Car drivers not following the correct procedure

Articulated lorries may have to adopt a different position to approach, circumnavigate and leave a roundabout due to their large size and length. The trailer will take a short cut and car drivers should take care not to become trapped in a lorry's manoeuvring space. Other drivers may not signal or position their vehicles correctly, and some may disregard the Give Way lines and enter the roundabout, causing other traffic to brake sharply.

Anticipating unexpected situations can help you to avoid becoming involved in accidents caused by the careless driving of others.

Crossroads

☐ Approach as for other junctions.

☐ Obey traffic light signals.

☐ Look both ways before emerging and give way if necessary.

☐ When turning right, pass offside to offside if possible, but nearside to nearside at staggered junctions or where road or traffic conditions dictate.

Turning right at crossroads needs care. It is safer to pass offside to offside. The disadvantage of passing nearside to nearside (that is, turning in front of oncoming traffic that is turning to their right) is that each car restricts the other's vision of oncoming traffic following them. Judgement of which manoeuvre to follow is dependent on your assessment of particular traffic situations. Look out for hidden hazards and be ready to stop if necessary.

Traffic lights

☐ Red means stop.

☐ Amber means stop.

☐ Green means you may go if it is clear.

In the case of an amber traffic light, keep going if you have already crossed the line or are so close to it that to stop would cause an accident. Always check the way is clear before driving on, and anticipate that lights may change or pedestrians may run across the road. Be prepared to give way to them.

Loads

Loading the car

☐ All loads must be absolutely secure.

☐ Ensure loads are spread as evenly as possible.

☐ Loads must not affect visibility or obscure lights.

☐ After loading, check headlights do not need re-adjusting.

It is an offence to drive with an unsafe load. Should anything fall from your car on a motorway, do not attempt to retrieve it yourself: stop on the hard shoulder and telephone for assistance. Spread loads so that the weight is not on one side of the vehicle. Luggage inside should be secure and not interfere with visibility or fastening of seat belts. Check headlight alignment after loading and unloading heavy loads.

Meeting traffic in narrow places

How to pass safely

☐ Give way to other traffic when it is their right of way.

☐ Give way to traffic if there is not room to pass safely.

☐ Do not squeeze past oncoming traffic.

☐ When giving way, wait well back and if

necessary, signal your intention to move out when safe.

Some roads narrow unexpectedly due to obstructions or parked vehicles, other older roads were not built with today's heavy traffic in mind. Look well ahead, anticipate problems and let approaching traffic pass if you are in doubt. Drive on when there is enough room to do so safely.

Passing cyclists or horses

☐ Slow in good time.

☐ When unable to pass safely, do not follow closely, allow plenty of room.

☐ Do not overtake until it is safe to do so.

☐ Pass as quietly as possible, without undue revving of the engine.

☐ Leave extra room after passing, before moving back to the left.

Cyclists can wobble or fall and horses can be startled by the noise or colour of a vehicle: treat both with great caution. Overtake, giving maximum clearance and, in case of animals, very slowly. When the cyclist or rider is travelling towards your vehicle, anticipate oncoming traffic may pass and, though it is *your* priority, slow and be ready to hang back to allow them through safely. A wrong decision on these occasions could result in loss of life and it can not be emphasised too strongly just how vunerable cyclists and horse riders are.

Mirrors - interior and exterior

Condition and limitations of mirrors

☐ Make sure mirrors are clean.

☐ Adjust the mirrors correctly before driving your vehicle.

☐ Remember mirrors have blind spots.

You should not have to move your head in order to see in the mirrors. Mirrors should be used frequently as a third eye and certainly before all driving actions. Outside mirrors may be made of convex glass to give a wider range of vision, and this will make it more difficult to judge the speed and distance of following vehicles and they should therefore be used in conjunction with the interior mirror. Drivers who have difficulty turning their neck or who have tunnel vision may be helped by fixing extra exterior mirrors. All mirrors have blind spots, and drivers should check alongside their vehicle before moving off or changing lanes.

Mirrors - signal - manoeuvre routine

Use before:

☐ any driving action;

☐ changing position or direction;

☐ overtaking;

☐ slowing;

☐ stopping.

The art of advanced driving is "never be taken by

surprise". In order to be aware, and anticipate the actions of others, good use of the mirrors is essential at all times. Awareness of blind spots is vital.

Motorway driving

Preparation for a motorway journey

☐   Do not travel if you are tired or unwell.

☐   Your car should be in a good mechanical condition.

☐   Check tyres for damage, wear and that the pressures are correct.

☐   Ensure fuel, oil and coolant levels are adequate for your journey.

☐   Study an up-to-date map and note your junction exit numbers.

☐   Plan regular stops for rest and refreshment at Service Areas.

☐   Listen to the local radio station for delays and diversions.

Motorways are the safest roads ever built. There are no roundabouts, crossroads, sharp bends, pedestrian crossings or very slow moving traffic. Vehicles are driven on a motorway travel at high speeds on carriageways of two or more lanes, making lane discipline and separation distances extremely important.

If you are travelling long distances at high speeds, remember to inflate your tyres 2 to 4 pounds per square inch (p.s.i.) above the recommended levels. See the manufacturer's handbook. Each year drivers are killed and cause the death of others by carrying out illegal manoeuvres. Obey the Highway Code: do not drive on motorways unless you are fully aware of the regulations.

Driving actions prohibited on motorways

- ☐ Stopping, unless in an emergency.
- ☐ Reversing.
- ☐ "U" turns.
- ☐ Crossing the central reservation.
- ☐ Driving against the traffic.

Joining a motorway

- ☐ Motorways often start at a roundabout.
- ☐ Blue rectangular signs give advance warning.
- ☐ Use the slip road leading to an acceleration lane if there is one.
- ☐ Give way to traffic already on the motorway.
- ☐ Keep in the left-hand lane until accustomed to the speed of the traffic.

The slip road joining a motorway leads to an acceleration lane. Use this lane to adjust your speed to that of the traffic already on the motorway. Enter the left-hand lane of the motorway when a suitable gap appears. When the traffic flow is heavy you may have to wait in the acceleration lane. Never force traffic to slow or swerve.

Motorway information signs

Comprise:

- [ ] advance warning signs of a motorway;
- [ ] start of the motorway;
- [ ] motorway interchanges;
- [ ] route confirmation signs after an exit;
- [ ] service areas;
- [ ] direction of nearest emergency telephone;
- [ ] motorway exit signs at approx. 1 mile and 1/2 mile;
- [ ] countdown markers to a deceleration lane at 300, 200 and 100 yards;

  end of motorway sign.

Motorway lights and studs

- [ ] Amber flashing lights mean hazard ahead, reduce speed to under 30 m.p.h..
- [ ] Red flashing lights mean STOP, do not go any further in that lane.
- [ ] Amber coloured studs mark the right-hand edge of the carriageway.
- [ ] Red coloured studs mark the left-hand edge of the carriageway.
- [ ] Green coloured studs separate the acceleration and deceleration lanes from the carriageway.

The flashing lights warn of dangers ahead, such as

fog or an accident. They will flash at intervals on the motorway, and some amber lights may have an illuminated sign giving advance warning of a lane closure, temporary advisory speed limits or the end of a previous restriction. If red lights flash on a slip road you must not enter it. The coloured studs are especially helpful at night when visibility is poor, and mark the edge of the carriageway.

Driving on a motorway

☐ Drive in the left-hand lane unless overtaking.

☐ Use dipped headlights in poor visibility or bad weather conditions.

☐ Leave an adequate distance between yourself and other vehicles.

☐ Use all mirrors constantly to monitor changing traffic situations.

☐ Drive in the middle of your lane.

Leave a gap *of at least* 1 metre per m.p.h. of your speed or follow the two second rule. A two second gap should be left between your car and the vehicle in front. Measure two seconds by repeating the phrase, "Only a fool breaks the two second rule". Obviously in bad weather conditions a longer gap must be left. Remember, heavy goods vehicles and those towing trailers are not allowed to use the third, right-hand lane.

Overtaking on motorways

☐ POSITION: well back from the vehicle you are about to overtake.

- [ ] SPEED: sufficient to complete the manoeuvre safely.

- [ ] LOOK: ahead for hazards and behind for traffic about to overtake or travelling in the lane you wish to enter.

- [ ] MIRRORS: use wing mirrors and check blind spots.

- [ ] SIGNAL: well before you pull out.

- [ ] MANOEUVRE: the car very gently at speed or you may lose control of the vehicle if you turn the steering wheel sharply. Do not move back into the left-hand lane until you are well past the vehicle being overtaken and can check this in the interior mirror and alongside.

Hazards to anticipate for before overtaking are: lane closures ahead; new traffic entering via an acceleration lane thus causing vehicles in front to change lanes; and traffic situations ahead that could affect your manoeuvre. Always check blind spots not covered by mirrors before pulling out, and ensure indicators are cancelled when the manoeuvre is completed. Self-cancelling mechanisms are not always activated by the gentle steering needed when overtaking at high speeds.

Night driving on motorways

- [ ] Use headlights, even when the carriageway is well lit.

- [ ] Use dipped headlights when there is oncoming traffic on the other carriageway or vehicles travelling in front of you.

- [ ] Do not wear tinted glasses or night

driving glasses.

☐ Extra care is needed before manoeuvres.

☐ Do not drive if tired.

Speed and distance can be deceptive at night, and great care needs to be taken when judging them. Use mirrors and signal earlier, allow more time for all manoeuvres. The lights from other traffic can be tiring and dazzling, so stop and rest at Service Areas more frequently.

Dealing with a breakdown on a motorway

☐ Pull onto the hard shoulder well away from moving traffic.

☐ Switch on hazard warning lights.

☐ If carried, place a warning reflective triangle 150 metres behind your vehicle.

☐ Do not allow passengers or animals to wander on or near the carriageway.

☐ Follow arrows on marker posts to nearest emergency telephone and ring for assistance. These calls are free but a charge will be made to remove your vehicle.

It is advisable to get passengers out of the vehicle but place them as far as possible from the carriageway, preferably over the fence in the next field. Statistics show more people are killed on the hard shoulder by vehicles colliding with them, than in actual accidents on motorways.

Advice for women travelling alone

- [ ] In the event of a breakdown pull over onto the hard shoulder, as far from the traffic as possible, and as near to an emergency telephone as you are able.

- [ ] Switch on hazard warning lights.

- [ ] If some distance from a phone, lock your door and wait in the car.

- [ ] If a member of the public offers assistance, do not open the door. Wind down the window an inch and ask them to contact the police or the breakdown service.

If you are able to stop next to an emergency telephone, call the police. If a member of the public stops, get back in your car, lock the doors and ask them to make the call for you.

This advice is offered in the light of recent attacks on women on motorways, and is purely advisory.

Leaving a motorway

- [ ] Use the advance warning signs to plan the distance to your exit.

- [ ] Move into the left-hand lane in good time.

- [ ] Use the mirrors, then signal on the approach to the first countdown marker.

- [ ] After the countdown markers move into the deceleration lane.

- [ ] Slow down in time for your exit road.

Move into the left-hand lane well before your exit and do not cross more than one lane at a time. Use mirrors and signal before changing lanes. Check the speedometer when you slow down in the deceleration lane. Many slip roads have sharp bends at the end of them, and after travelling some distance at high speeds your judgement will be affected, resulting in you thinking that you have slowed more than you actually have.

Night driving

Lights at night

☐ Check lights are working properly.

☐ Keep lights clean and adjusted correctly

☐ Switch headlights on as soon as daylight begins to fade.

☐ Dip headlights when necessary.

☐ You may park on a road subject to a 30 m.p.h. speed limit without lights.

Lights fulfil two functions: they enable a driver to see and be seen. It is an offence to drive a vehicle if the lights are not working properly. Replace bulbs immediately they fail to work. Incorrectly adjusted headlights can dazzle other drivers. Be the first to switch headlights on at lighting up time, or earlier if the light is bad. Dip headlights in a built-up area and for oncoming traffic or when there is traffic travelling in front of you. Cars travelling into a left-hand bend need to dip much earlier than cars entering a right-hand bend if there is oncoming traffic.

Speed at night

☐ Drive at a speed at which you can stop safely within the distance illuminated by your lights.

☐ Take care before judging the speed of oncoming traffic at night, lights do not convey an accurate impression of speed.

☐ Slow down sufficently before entering bends.

At night, speed and distance can be deceptive, as many drivers have found to their cost. On dark nights, especially when it is raining, visibility can be as poor as in foggy conditions. Never wear tinted lenses or so-called night driving glasses (see the section on eyesight for more information).

Use of the horn at night

☐ It is an offence to use the horn at night between 11.30p.m. and 7.00 a.m. in built-up areas.

Overtaking

When to overtake

☐ When it is really necessary.

☐ When legal.

☐ When you can complete the manoeuvre safely.

When it is illegal to overtake

☐ Where there is a no overtaking sign

☐ Where there are double white lines and it would mean crossing a solid line on your side of the road.

☐ Within the zigzag lines approaching a pedestrian or pelican crossing.

You may cross a solid white line to pass parked vehicles, turn into an entrance or if signalled to do so by a policeman or traffic controller, but take care.

Overtaking safely

☐ Allow an adequate distance between your vehicle and the one you intend to overtake.

☐ Ensure your speed is sufficient to complete the manoeuvre safely.

☐ Check it is clear ahead and behind before pulling out.

☐ Allow a good clearance when passing other vehicles.

☐ Check in mirrors and alongside before moving back in. Do not cut in front of vehicles.

Always leave plenty of room for this manoeuvre. 80 per cent of all accidents happen to vehicles that are overtaking or turning right. When passing cyclists or motorcyclists, leave enough room that if they were to wobble or fall they would be in no danger from your vehicle. If in doubt whether to overtake – DON'T. Remember, *overtakers get to the undertakers quicker!*

Parking a vehicle

Where to park

☐ Park in a legal place.

☐ Park in a safe place.

☐ Park in a convenient place.

Common sense will tell you whether the place you have chosen is safe and convenient. For example, avoid narrow roads, bends, entrances and drive-ways. The Highway Code details where it is illegal to park and mentions motorways; sections of roads where there are double white lines marked along the centre; and within the zigzag lines on either side of a pedestrian crossing among others.

Parking at night

☐ Park on the left-hand side of the road, unless in a one way street.

☐ Parked cars must show lights unless in a built up area.

☐ Do not park within 15 metres of a road junction.

Passenger vehicles may park on a road with a 30 m.p.h. speed limit without lights, provided they have no overhanging or projecting load or a trailer attached. Vehicles must be parked close and paral-lel to the kerb. On roads not subject to a 30 m.p.h. speed limit, both front and rear lights must be shown on parked cars. One parking light or offside and rear light is no longer legal.

Parking in a car park

☐ Follow any arrows or one way systems that are in use.

☐ Choose a large enough space to allow doors to open safely.

☐ Use dipped headlights in indoor or multi-storey car parks in poor visibility.

☐ Reverse into parking spaces where possible.

Take care when parking in car parks, especially when they are busy. Make good use of mirrors and watch for other cars, children and shoppers who may be unaware your vehicle is moving. It is normally safer and easier to manoeuvre by reversing into a parking space and driving out. In multi-storey car parks keep the parking-ticket, on your person. This will make it more difficult for car thieves to leave the car park with your vehicle. Do leave room for doors to open when you park, it may be more difficult to pursue insurance claims for damages when not parked on a public highway.

Parking on a road between two vehicles

☐ Choose a large enough space.

☐ Reverse in, unless there is plenty of room to drive in.

☐ When reversing, drive half a length past the leading vehicle and stop.

☐ Remember the front of your car will swing out as you steer.

☐ Constant observation is needed.

☐ Give way to pedestrians and other traffic.

It is unwise to choose a small space. The drivers of the cars parked either side of you may not have your expertise and could damage your vehicle when they drive off. To reverse in, you need at least one and a half car lengths, preferably two, and a much larger space is needed to drive in forwards.

When reversing, drive past the gap and stop about 1 metre away from, and parallel to, the leading vehicle. Reverse until your rear wheels are roughly level with the rear wheels of the parked vehicle, then turn the steering wheel slightly to the left. Check round for pedestrians and other traffic, the front of the vehicle will swing out at this point. When your front wheel is roughly level with the other vehicle's back wheel, begin to straighten up. Make sure the front nearside corner of your car clears the corner of the other car. When you are clear of the other car and your offside wing-mirror lines up with the near-side headlight of the car behind you, quickly apply right lock. When in the parking place, stop and straighten up as necessary.

It is dangerous to park on a road in foggy conditions

Parking on a gradient

☐ Leave the car in gear to prevent rolling should the handbrake fail.

☐ Turn the wheels into the kerb as an additional precaution.

☐ Make sure the handbrake is firmly applied.

When parking uphill, leave the car in first gear with the steering wheel turned to the right. If the car should roll it will be stopped by the front wheel coming into contact with the kerb. When parking downhill, leave the car in reverse gear and turn the steering wheel to the left.

Parking on soft ground

When it is necessary to leave the road to park or turn on soft ground, keep the driving wheels on a hard surface to prevent sinking and wheel spin.

Passing parked vehicles

Passing parked vehicles safely

☐ Look ahead and decide if it is safe to pass the parked vehicle.

☐ If there is not room to pass safely, wait well back.

☐ Signal your intention to pull out if necessary.

☐ Give a good clearance in case of unexpected hazards.

☐ Do not pull in until there is sufficient room to do so safely.

If you have to wait, wait well back, turning the front wheels slightly to the right to aid steering when moving off. When passing the vehicle, anticipate a door may open, a child may run out or the vehicle begin to move off. Check in mirrors and alongside before pulling back in. Where there are several parked vehicles and it is safe, keep out until you have passed all the obstructions, do not weave in and out unless necessary to wait for more oncoming traffic.

Pedestrian and pelican crossings

Pedestrian crossings

☐ Look ahead for advanced warning signs of crossings.

☐ Stop and give way to people on a crossing.

☐ Slow and be prepared to stop for people waiting to cross.

☐ Do not park on or near the zigzag lines of a pedestrian crossing.

☐ Do not overtake within the zigzag lines approaching a pedestrian crossing.

Pedestrians using a pedestrian crossing have an absolute right of way and drivers must stop and wait until they have finished crossing. Where there is a central refuge treat both halves as a separate crossing. An arm signal will give pedestrians confirmation you are slowing down or stopping.

Pelican (pedestrian-controlled) crossings

☐ Look ahead for the advanced warning signs.

☐ STOP if the lights are red.

☐ STOP if people are crossing when the lights are flashing amber.

☐ STOP when the lights are amber, unless to do so would cause an accident.

☐ Green means GO if clear.

☐ Zigzag lines are now being placed on the approach to pelican crossings; do not park or overtake within these zigzag lines.

Give way to pedestrians who are crossing. Treat a pelican crossing that is staggered with a central refuge, as two separate crossings. Watch out for people pressing the button on a pedestrian-controlled crossing, as the lights change quickly and it is wise to anticipate this and slow in good time.

Reversing

When it is illegal to reverse

☐ On a motorway

☐ From a side road into a main road.

☐ Longer than necessary.

☐ The wrong way up a one-way street.

Make sure it is safe before you reverse and ensure reversing lights, if fitted, are working. There is a tendency for some people to reverse out of driveways into main roads: it is far safer to reverse in, and drive out. When you reverse, ensure the manoeuvre is safe, legal and does not inconvenience other motorists.

Reversing safety points

☐ Good all-round observation is essential

☐ Do not rely on mirrors, but check blind spots.

☐ The front of your vehicle may swing out as you turn causing danger to others.

☐ Check for people, especially small children crossing or standing behind your vehicle, and *Give Way* to them.

Each year small children are killed by reversing vehicles, some, tragically, by their parents. If in doubt get out of your vehicle and check round the car before reversing. When drivers reverse they are allowed by law to remove their seat belt. Turn round and look where you are going: give way to other vehicles approaching from behind when you reverse.

Tips for reversing

☐ The car takes longer to respond to steering.

☐ The brakes can be less responsive on a reversing vehicle.

☐ If passengers are blocking your view, ask them to move.

☐ Small drivers may find altering the seat or using a cushion helpful to increase their range of vision when reversing.

The golden rules for reversing are, take care and take time. If you misjudge a reverse, pull forward and start again.

Seat belts / child safety seats / carry cot restraints

Condition and fixing of restraints

☐ Make sure seat belts are not damaged or frayed.

☐ Anchorage points must not be rusted.

☐ Fixing bolts must be properly secured.

☐ Buckles should fasten and release easily.

☐ Check the belt locks when it is pulled suddenly.

All the above is pointless if you drive off without putting your seat belt on. Unless in a special category, all drivers and front seat passengers must wear a seat belt when the car is in motion. Do not be tempted to leave your seat belt off for short journeys, statistically this is when accidents are likely to occur and you will need the protection a seat belt provides. You may remove the seat belt when reversing.

Back seat passengers

☐ If the back seats are fitted with seat belts, ask passengers sitting in the rear to wear them.

☐ The law requires to be restrained where belts are fitted.

☐ Cushions anchored to the seats can make seatbelts safer and more comfortable for small children.

Sitting in the back seat does not protect passengers from injury in a sudden stop or accident. Front seat occupants can be killed or seriously injured by rear seat passengers flying forward. If belts are fitted ensure they are worn by back seat passengers.

Never leave children alone in a vehicle.

Child safety seats

☐ There are special backward facing seats for babies of up to 9 months.

☐ Child safety seats can be used by children between 6 months to 4 years.

☐ Junior safety harnesses are made for children aged 4 to 11 years.

There are a variety of safety seats on the market and they can be fitted to the front or rear seats either using anchorage points or restrained in adult seat belts. Make sure any seat is British Standards Approved and fitted securely according to the maker's instructions. Insist they are used at ALL times by children travelling in the car.

Carry cot restraints

☐ Babies in carry cots should be strapped securely in the cot.

☐ Carry cots should be secured by a properly fitted carry cot restraint on the rear seat of the car.

Seating

How to adjust your seat

☐ Ensure your feet can reach the foot pedals comfortably.

☐ Make sure the seat is locked into position and will not slide.

☐ Adjust the back of the seat so you can hold the steering wheel with arms slightly bent.

☐ Head restraints must not be below eye level for maximum protection.

Unfortunately most cars are designed for drivers of an average height and weight. If you do not come into this category it may be more difficult to adjust your seating. Smaller people may find a strategically placed cushion helpful. Do not be tempted to remove head restraints to aid visibility. They are not provided as head rests but to prevent the neck from whiplash and other serious injury in the event of an accident.

Separation distances

Judging separation distances

☐ Know your stopping distances: they are in the Highway Code.

☐ Follow the two second rule.

☐ Make allowances for poor weather conditions and road surfaces. Keep well back from vehicles throwing up spray.

☐ Let drivers following closely pass as soon as it is safe.

Do not follow traffic closer than your overall stopping distance. Leave sufficient space to stop safely should the vehicles in front stop suddenly. Allow double the distance in wet weather and take extra care in foggy or icy conditions to reduce speed according to the severity of the road conditions. On the open road on a dry day, a two second gap or a gap of 1 metre for each m.p.h. of your speed should be sufficient. On motorways, when travelling at 70 m.p.h. in good conditions, keep the equivalent of the distance between two marker posts behind the vehicle in front.

Gently slow if safe to do so, to allow drivers who are following too closely to pass .

Signals

When to give arm signals

☐ If the indicators stop working, perhaps due to unexpected bulb failure.

☐ To signal an intention to turn right after passing a parked vehicle.

☐ When slowing or stopping at a pedestrian crossing.

☐ To help other road users.

When arm signals are given, they should be given correctly and clearly or they will be misunderstood. Flapping a hand out of the window may confuse another driver, extend your arm fully and make sure other drivers have noted your signal.

When to indicate

☐ When moving off.

☐ When changing position, direction, or overtaking.

☐ When stopping on the left or right.

Common sense should be used as to whether a signal is necessary or not. Generally speaking, always give a signal when it is needed to warn any other road user of your intention to change position. It is not always necessary to signal before passing parked vehicles if the driver behind is aware of your impending manoeuvre.

Should a vehicle be parked on a bend, or if you have to wait back prior to passing parked vehicles, a signal could well be necessary. Drivers do not need to signal left as a matter of course when completing an overtaking manoeuvre, but should signal in good time when changing lanes on a motorway. Signal your intentions in plenty of time. Cancel your signal when the manoeuvre is completed to avoid misleading others.

Signals given to or by others

☐ Do not beckon pedestrians or other drivers. Let them make their own decisions, as you may wave them into danger.

☐ Do not flash your lights at others unless to warn them.

☐ When others signal to you, act only if it is safe to do so.

Apart from warning there is no recognised meaning of flashing headlights, you may confuse other drivers or place them in danger by "flashing them on".

Skidding and aquaplaning

Causes of skidding (An *ABC*)

☐ *A*ccelerating sharply.

☐ *B*raking sharply.

☐ *C*ornering sharply.

Skidding is not caused by slippery road conditions, but by drivers not predicting or reacting to them. Take care if the steering feels "light" indicating the tyres are losing their grip on the road surface.

Factors which affect road holding

☐ Wet roads.

☐ Icy or snowy conditions.

☐ Loose road surfaces.

☐ Unbalanced brakes.

☐ Worn or wrongly mixed tyres.

Wet roads can be several times more slippery than dry ones and at high speeds water can build up under the tyres causing the car to slide or aquaplane. You can double your grip by reducing your speed by half. It will take at least twice as long to stop in wet weather. Wet roads may be muddy, and wet mud can also be as slippery as ice. Watch for loose road surfaces which may be caused by gravel or fallen leaves.

Make sure your brakes work evenly on all four wheels, or if you brake hard on a wet surface the car could spin. Tyres should be inspected regularly for damage and a good tread depth. The legal depth is 1 millimetre, but it is unwise to allow the tread to wear below 3 millimetres if you wish to stop safely in bad weather conditions. It is illegal to mix radial and cross-ply tyres unless there are two cross-ply tyres on the front axle and two radial on the back, but this is not recommended. See page 73 for further information about tyres.

How to avoid skidding

☐ Maintain your brakes properly.

☐ Regularly check the tread depth of your tyres.

☐ Accelerate, brake and corner gently.

☐ Watch for changes in road surfaces.

☐ Drive at a safe speed according to the conditions.

The most important point to remember is to drive slowly when the road is slippery, to look ahead and anticipate hazards. Accelerate gently, brake on the straight not on bends and avoid turning the steering wheel sharply. After rain, grease and oil float on the surface of the water creating an additional hazard on roads.

How to deal with a skid

☐ If the driving wheels are spinning, decelerate until the wheels start gripping the road surface.

☐ In any skid, *do not brake: leave pedals alone*

☐ In a rear wheel skid, turn into the skid.

☐ In a front wheel skid, decelerate do not steer.

Skids can not be corrected but the effects can be minimised if you deal with them properly and avoid braking, accelerating or steering. The only exception is in a rear wheel skid: when the back starts to slide, steer gently in the direction that the back of the car is sliding but beware of oversteering, it could lead to a skid in the other direction.

How to avoid aquaplaning

☐ Reduce speed to below 50 m.p.h. in wet weather.

☐ Ensure all tyres have adequate tread: a *minimum* of 1 millimetre, but preferably 3 millimetres.

When tyres are turning in wet weather, a ridge of water builds underneath them. The higher the speed, the more water, so there can be a film of water between the tyres and the road. The driver may be unaware of the problem until altering speed or changing direction, the car will be virtually without steering or brakes. Whether an accident occurs depends on the traffic situation and luck!

Snow and ice

How icy conditions affect vehicles

☐ Windows and mirrors may ice up: keep them clear.

- ☐ Remove snow from indicators and lights.

- ☐ Stopping distances may increase *ten* times or more.

- ☐ Snow can pack under the wheel arches and round tyres and ultimately affect braking and steering.

- ☐ When very cold, leave the handbrake off at night to avoid it freezing on.

- ☐ Park facing away from the wind when snow is expected. This will protect your engine from drifting snow.

In snow it may be necessary to stop from time to time to clear snow away from windows and lights. When heavy snow is falling, stay off the roads unless your journey is absolutely essential. Listen to local radio stations, or contact one of the motoring organisations for up to date information of weather conditions. If you have to leave your car in the open, due to heavy snow or drifting, tie something bright to the aerial to assist with finding your vehicle.

Moving off in snow

- ☐ Keep some sacking in the car to place under the driving wheels.

- ☐ Use the highest gear you can (probably 2nd gear), to move off.

- ☐ Keep a heavy object in the boot of rear-wheel drive vehicles.

- ☐ Keep a spade in the boot to dig snow away if necessary.

Driving in snow or ice

☐ Light steering or lack of tyre noise may warn of black ice which is not always readily visible.

☐ Operate all driving controls gently.

☐ Select a lower gear earlier before travelling downhill.

☐ Select the best gear to climb uphill and stay in it.

☐ Keep warm and waterproof clothing handy in case you are forced to walk.

In slippery conditions drive very carefully, making more use of engine braking and avoid using the footbrake, other than very gently. Do not brake and steer at the same time.

Speeds and speed limits

National speed limits for cars

☐ 70 m.p.h. on motorways.

☐ 70 m.p.h. on dual carriageways.

☐ 60 m.p.h. on single carriageways unless signs show otherwise.

☐ 30 m.p.h. in built-up areas where there is street lighting, unless signs show to the contrary.

Remember these are the maximum and legally enforceable speed limits allowed on roads in the United Kingdom. In many cases it may be both unwise and unsafe to keep up to them. Blue circular signs with a speed limit on them, indicate a minimum speed and are likely to be found where there are long underpasses or tunnels. Take care driving at night when speed can be dangerously deceptive and reduce speed when traffic or weather conditions are bad.

Steering

Correct steering habits

- ☐ Make sure you are seated comfortably.

- ☐ Hold the steering wheel correctly.

- ☐ Feed the wheel round with a pull and push movement: do not cross your hands.

- ☐ Do not allow the steering wheel to spin back after turning.

Bad steering habits are quickly adopted by many drivers after passing their driving test. Driving with only one hand on the wheel and crossing hands when turning the wheel are the most common faults. Hold the wheel at the ten to two, or quarter to three position for normal driving. This will ensure that should an emergency arise, you will have maximum control at all times. Drivers who drive with one hand resting on the gear lever are actually causing undue wear on the gear selector rods.

Stopping quickly in an emergency

Basic rules

☐ Hold the steering wheel firmly.

☐ Brake quickly but progessively.

☐ Leave the clutch alone until just before you stop.

☐ When you have stopped apply the handbrake, change the gear to neutral and switch off the engine.

It is necessary to hold the wheel firmly in order to avoid swerving. If you brake too harshly, the wheels will lock causing the car to skid. Should this happen, release the foot brake, allow the wheels to start turning again, then apply the brakes more gently. Do not depress the clutch until the vehicle has almost stopped in order to use the stability that engine braking gives to the steering and braking systems of the car.

Using cadence braking

☐ Hold the steering wheel firmly.

☐ Apply the brake pedal sharply, release it before the wheels lock and repeat again and again until the car has stopped safely.

The brakes are most effective just before the wheels lock. Anti-locking braking systems fitted to some cars automatically apply and release the brakes in such a way that the vehicle stops quickly without losing control. Cadence braking is not as effective as these systems but with practice drivers should be able to stop under control in slippery conditions. The harder the brakes are pumped, the more effective the braking. It is advisable to practice this technique in a safe place.

Factors which affect stopping

- ☐ Speed at which the vehicle is travelling.

- ☐ Condition of brakes, shock absorbers and tyres.

- ☐ Surface of road.

- ☐ Weather conditions.

- ☐ How promptly the driver reacts and begins to apply the brakes.

Even modern technology cannot instantly stop a vehicle which is travelling at speed. It will take an alert driver about the same distance in feet as the speed in m.p.h. that his vehicle is travelling, to react and begin to apply the footbrake. At 30 m.p.h. on a dry road, in a car with effective brakes and tyres in good condition, an alert driver will be able to stop a vehicle in approximately 75 feet or 23 metres. At 70 m.p.h. the overall stopping distance will be 315 feet or 96 metres. Drivers should always make allowances for poor weather conditions, as it will take at least twice as long to stop in wet weather and probably ten times as long to stop in snowy or icy conditions.

Towing a trailer or caravan

Rules for safe towing.

☐ The trailer attachment should not be worn.

☐ The trailer should be securely attached to the towing vehicle including the safety chain.

☐ Check the rear lights and indicators on the back of the trailer or caravan are in good working order and connected properly.

☐ Comply with the national speed limits for vehicles towing trailers.

☐ Fit extended mirrors where necessary.

☐ Do not use the third, right-hand lane on a motorway.

Cars towing caravans or trailers are allowed to travel at a maximum speed of 60 m.p.h. on a motorway or dual carriageway, and 50 m.p.h. on a single carriageway where national speed limits apply. Extra care should be taken, and speed reduced, when there are high winds in exposed places. Caravans should never be towed when occupied.

Tyres

Legal requirements for tyres

☐ At least 1 mm of tread covering the circumference and three-quarters of the width of the tyre.

☐ Tyres must be inflated to the correct pressure.

- ☐ No cuts, bulges, breaks in fabric or ply structure exposed.

- ☐ Do not mix radial and cross-ply tyres on the same axle.

- ☐ If tyres are mixed, the two radials must be fitted on the rear.

- ☐ A spare tyre, if carried, must comply with legal requirements.

Though 1 mm is the legal minimum depth of tread allowed on tyres, it is inadvisable to allow tread to wear below 3 mm if you want good road holding in wet weather. Correctly inflated tyres will last longer. Check pressures when tyres are cold and remember many gauges are inaccurate, so double check with another gauge if possible. Do not confuse metric and imperial measurements on gauges.

If driving at speed for long distances increase normal tyre pressures by 2 to 4 p.s.i. on each tyre or as recommended by the manufacturer. Before driving your vehicle, walk round it and have a good look at the tyres, especially for damage that may have occured to the walls of the tyre. It is not advisable to mix tyres, it is safer to have the same type of tyres on all the wheels including the spare wheel. If a mixture is unavoidable, remember *radials to the rear.*

Tyre safety rules

- ☐ Fit tyres of the size and type recommended by the manufacturer of your vehicle.

- ☐ Check tyres regularly and replace when necessary.

- ☐ Have your wheels balanced and the tracking checked regularly.

☐ Keep tyres correctly inflated and replace valve caps.

☐ Any repairs must be done professionally by a vulcanising process.

Check your tyres frequently and try to avoid damaging them against kerbs, in potholes or against obstructions. Avoid cornering, accelerating or braking harshly. Skidding can dramatically reduce the life of tyres. Have repaired tyres fitted to the rear of the vehicle and never use tyres that have been temporarily repaired with a plug, other than to get you home, and then drive below 40 m.p.h.. New tyres should be bedded in at low speeds for the first 100 miles. Wash tyres regularly in the winter as prologed exposure to salt can damage them.

Worn tyres can kill and maim.

Burst tyres

☐ Hold the steering wheel firmly.

☐ Do not attempt to brake.

☐ Allow the car to roll to a halt if possible.

Prevention is better than cure so check your tyres frequently. If in doubt have a tyre examined professionally or replaced. It is likely a car will swerve when a tyre bursts so leave the brakes alone and try to roll to a safe stop.

Tyres are the only contact between your car and the road...look after them.

Windscreens and windows

Keeping windows clear and clean

☐ Keep windscreens and windows clean.

☐ Do not reduce visibility with unnecessary stickers or hanging soft toys.

☐ Replace cracked or badly scratched windscreens.

☐ Do not allow windows to mist up, use demisters in good time.

If the above rules are not followed visibility will be reduced increasing the chance of the driver's eyes missing a vital piece of information when making a driving decision, and thus be involved in an accident. Ensure all windows are completely clear before driving in icy or snowy conditions.

Windscreen dazzle

Drivers may occasionally be dazzled by headlights. You should not look at the oncoming vehicle, but slow down and stop if necessary. The winter sun can sometimes blind drivers, bathing the view ahead in a bright orange glow. Rather like driving in a thick fog, slow down in order to stop within your range of vision.

Avoiding and dealing with shattered windscreen

☐ Drive more slowly on loose surfaces.

☐ Do not follow traffic closely, and allow extra separation distance on loose surfaces.

- [ ] In the event of a shattered windscreen try to stop in a safe place.

- [ ] Cover body work and air vents of car.

- [ ] Remove as much glass as possible by pushing outwards with protected hands.

- [ ] *If necessary to drive*, drive slowly to minimize the risk of flying glass. Replace the windscreen as soon as possible.

Conclusion: Advanced driving tests

Motorists who would like an expert opinion on their driving skills can contact a professional driving instructor or one of the road safety organisations which specialize in conducting advanced driving tests.

Contact one of the following for information:

Driving Instructors Association Ltd.
Lion Green Road,
Coulsdon,
Surrey CR3 2NL.
Tel: 01 - 660 3333

They will be pleased to supply details of highly qualified driving instructors in your area, who will be able to assess your driving and give advanced tuition as required.

Institute of Advanced Motorists
359 Chiswick High Road,
London W4 4HS.
Tel: 01 - 994 4403

All examiners have the Advanced Police Certificate and local organisations offer training courses.

Royal Society for Prevention of Accidents (RoSPA)
Advanced Drivers Association,
Cannon House,
The Priory Queensway,
Birmingham B4 6BS.

All examiners are Police Class 1 drivers and candidates who pass are required to sign a code of conduct.

Books by the same authors

Driving Test - Questions and Answers
Robbins and Robbins

Written by two practising instructors this booklet is designed to fill the gap that many learner drivers feel when they encounter the literature currently available. The text is a draft of a number of questions most commonly asked during driving tests. Answers are situated on the facing page.

How to Pass Your Driving Test
Robbins and Robbins

Once again the authors have attempted to fill a gap in the available literature on the driving test. This booklet includes detailes information explaining how an examiner conducts and marks a driving test and the standards of driving required to pass.

Other Books of Interest

Test Yourself on the Highway Code
Geoffrey Whitehead

Learner drivers will find this a boom in helping them understand and absorb the rules in the Highway Code. Consisting of 400 self-test questions, with diagrams and illustrations, it is designed so that answers to questions which appear on the same page, can be concealed whilst the questions are read.

Don't Get Ripped Off

Rawlinson

The author goes into great detail regarding what to look for when purchasing a second-hand car. This most useful book is enhanced by a 102-point check list covering everything from antifreeze to windscreen wipers.